THE SIMPLE SORCERER'S ILLUSTRATED PROBABILITY PRIMER

written by Lin Sten
illustrated by David Gonzales

Text and illustrations copyright © 2004 by Lin Sten

© 2004 Sten

All rights reserved. No part of this publication may be reproduced, stored in a retrieval system, or transmitted in any form or by any means, electronic, mechanical, optical, photocopying, recording, scanning, or otherwise, without written permission of the copyright holder.

For information regarding permission, contact the copyright holder:

Lin Sten
c/o. Custom Books Publishing
CreateSpace
100 Enterprise Way, Suite A200
Scotts Valley, CA 95066
USA

Library of Congress Cataloging in Publication Data:
Sten, Lin
The simple sorcerer's illustrated probability primer/ Lin Sten
Includes index.

ISBN 1434839923
EAN-13 9781434839923

**The Simple Sorcerer's
Illustrated
Probability Primer**

Written by Lin Sten
Illustrated by David Gonzales

This book is for anyone who might have a present or future interest in probability.

Early familiarization with symbols and their use is helpful to their later understanding; a child does not have to initially understand the symbols to benefit from their introduction. Perhaps the best known example of our belief of this is in the wooden alphabet blocks that almost every child receives before he or she can talk.

Any adult—teacher or interested parent—can read the text and captions for a fuller understanding while explaining as necessary to a child who might at first only enjoy the color panel illustrations, which are coordinated with the text on the facing page.

Any student of science, statistics, and engineering will find this book useful because it contains all the basic formulas of probability, gives a simple review of the fundamentals of probability, and offers both simple modeling examples and examples of practical applications.

In the illustrations there are seven characters who appear:

The Sorcerer

Aprendiza—the sorcerer's apprentice

Pet—the petulant probability figure

Owl—a mascot symbolizing the wisdom of knowing about probability

Farmer

Franco—the farmer's assistant

Dukelis—the donkey

The Simple Sorcerer's
ILLUSTRATED PROBABILITY PRIMER

Table of Contents

Preface	v
Activities in the Everyday World	ix
Related Activities in the Modeling World	x
Introduction	xi
Simple Things to Know or Learn	xiii
Procedures and Outcomes	1
Outcomes and Sample Spaces	3
Events on a Sample Space	5
Lists, Statements, and the Meaning of Events	7
Simple Probability	9
Simple Probability Examples	11
Event Operations and Venn Diagrams	13
Complements Rule	15
Addition Rule	17
Conditional Probability	19
Product Rule	21
Decision Trees	23
Practical Applications	25
Glossary	32
Appendix	35
Index	37

Preface

This book offers an introduction to probability that is enjoyable and educational for both adults and children. Each chapter consists of one page of text and one page of illustrations.

This book was created with the goal of presenting a limited amount of material in a manner that almost anyone can understand. Thus, it was made simple without having to falsely claim that the subject of probability can be made simple.

Based on the material that precedes it, the last chapter contains several practical examples, which show the importance of probability in the real world.

An appendix is attached for those who need a refresher on fractions or who wish for more details on a topic. This is followed by the index.

After the index is a section that offers some comprehensive problems for the advanced student.

Aerial view

The castle across the stream from the farm

Activities in the Everyday World

For activities in the everyday world that can be profitably analyzed with probability, we can imagine any of the following:

 horse racing

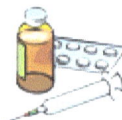 medicine (batch testing)

 manufacturing (batch testing; quality control)

 agriculture

 gambling

 lottery

  stock market (investment)

 judicial decision

 insurance

Related Activities in the Modeling World

For activities in the mathematical modeling world that can be analyzed with probability, and that help us better understand how to use probability in the everyday world, we can use or imagine any of the following:

toss a coin

roll 1 die

randomly draw a marble from an urn

roll 2 dice

flick a spinner

These modeling ideas are closely related to the study of probability, where problems are idealized for the clarity of the discussion and learning.

Introduction

Probability is magic,

                and some effort is necessary to learn the craft.

The methods, language, and symbols of probability allow us to quantify our partial knowledge about the world.

Probability is always dependent on point of view. What some other person, animal, alien, or god may know about the past, present, or future is irrelevant to what you compute based on your present knowledge.

Simple Things
to
Know or Learn

Faith helps you have patience with yourself while you learn something that is certain to change your life.

Symbols have value. Historically words, which are verbal symbols, first spoken and later written, were used in ancient Egyptian, Sumerian, and Chinese civilizations to express both romantic and mathematical ideas, were associated with the gods, and were thought to have magical power. This is no wonder, though we usually take this power for granted today, because words expand your mind, giving you the ability to articulate ideas, increasing your ability to organize thoughts, and aiding your communication with other people. Mathematical symbols extend this power even further.

Logic enhanced our survival even before we had words or other symbols. Primitive logic can exist and have meaning without articulation: "There has been water in the creek every day that I remember, so I will seek water there now." This idea (inductive reasoning) can exist in animals and it existed in humans before there were words that allowed its articulation, and long ago it guided us to water.

Fractions are fundamental to probability. (See the appendix.)

Procedures and Outcomes

The idea of probability is based on carrying out, conducting, exercising, doing, or running a *procedure*.

A *procedure* is a specification for an action and an observation.

The possible *outcomes* of a procedure is the list of possible results (real or imagined) given that we follow the procedure. These possible outcomes are always of interest if we wish to organize our thoughts.

Simple examples follow.

Procedure #1: Roll a 4-sided die once, and observe the number.
The possible outcomes are 1, 2, 3, and 4.

Procedure #2: Roll a 4-sided die once, and observe whether the number is even or odd.
The possible outcomes are even and odd.

Procedure #3: Toss a coin once, and observe whether heads (h) or tails (t) shows.
The possible outcomes are h and t.

Procedure #4: Toss a coin once, and observe whether heads (h) or tails (t) shows or if it lands on its edge (e).
The possible outcomes are h, t, and e.

Procedure #5: Toss a coin once, and observe whether it lands on its edge (e) or not (ne).
The possible outcomes are e and ne.

Procedure #6: (Randomly) draw one marble from an urn containing 1 blue (b) marble, 2 identical green (g) marbles, and 5 identical red (r) marbles, and observe the color.
The possible outcomes are blue (b), green (g), and red (r).

Procedure #7: (Randomly) draw one marble from an urn containing 1 blue marble, 2 identical green marbles, and 5 identical red marbles, and observe whether it is blue (b) or not (nb).
The possible outcomes are b and nb.

Remember that only one outcome results from one run of the procedure, no matter how many or how few possible outcomes there are.

Additional procedures are defined in Decision Trees and in Practical Applications.

Outcomes and Sample Spaces

In a single run of a procedure there will be only one outcome, though there might be many possible outcomes. We simplify and formalize what is possible by writing a symbolic list of all possible outcomes. This list is called the *sample space* (for the given procedure); for it we use the symbol S.

The symbol "$\equiv$" is read "is equivalent to", and it is similar to equality. (We will restrict the concept of equality, the use of the word "equals" ("="), to numerical relationships; "$\equiv$" will be used to relate what we will call equivalent events, which will be discussed later.)

Simple examples follow.

For Procedure #1, S $\equiv$ { 1, 2, 3, 4} $\equiv$ {2, 3, 1, 4}, etc.
(order is irrelevant among the possible outcomes in S)

For Procedure #2, S $\equiv$ {even, odd}

For Procedure #3, S $\equiv$ {h, t}

For Procedure #4, S $\equiv$ {h, t, e}

For Procedure #5, S $\equiv$ {e, ne}

For Procedure #6, S $\equiv$ {b, g, r} $\equiv$ {r, g, b} , etc.

For Procedure #7, S $\equiv$ {b, nb}

Remember that no matter how many or how few possible outcomes there are in the sample space, only one outcome can result from a single run of the procedure.

Additional examples are given in Decision Trees and in Practical Applications.

Events on a Sample Space

The formality of the complete symbolic list of possible outcomes for a procedure, giving us the sample space for the procedure, can be fully exploited if we are willing to be as formal with partial lists and their meaning. Any list that contains only possible outcomes from a given procedure is called an *event* on the corresponding sample space.

A simple example follows.

For Procedure #6, S ≡ {r, g, b},
and there are eight events on S:

S itself is an event, and it is called the *certain event*.

{r, g}
{r, b}   These first four events are *compound events*.
{g, b}

{r}
{g}   These events with only one outcome are *simple events*.
{b}

{ }   This is called the *impossible event*;
it contains no outcomes.

Can you write down all of the events on S for Procedure #3? (There are four events on S.)

Can you write down all of the events on S for Procedure #4? Since there are three possible outcomes in S, there are eight events on this sample space.

Remember that only one outcome results from one run of the procedure. No matter how many or how few possible outcomes there are in the sample space, or how many events there are on the sample space, only one outcome can result from a single run of the procedure.

Additional examples are given in Decision Trees and in Practical Applications.

Lists, Statements, and the Meaning of Events

The thought of a specific procedure gives rise to all its possible outcomes, the corresponding sample space, and to all the events on that sample space. Nonetheless, a (single) run of the procedure results in only one outcome.

As a simple example,
consider Procedure #6, for which $S \equiv \{r, g, b\}$.

Let us do one run. Suppose we draw a blue marble, i.e., the outcome is blue.

Then we could say all the following events *occurred*:
$\{b\}$, $\{r, b\}$, $\{g, b\}$, S (the certain event)

These events *did not occur*:
$\{r\}$, $\{g\}$, $\{r, g\}$, $\{\ \}$ (the impossible event)

Despite the power of these symbols, every practical probability problem must connect to symbols through statements, which are (define) events. It is common to use present tense ("is") in these statements while we imagine a future time at which the statement becomes true or false when the event occurs or not.

On the sample space S for Procedure #6, we have this event for example:
        a randomly drawn marble is red or blue.
It is *equivalent* to ("$\equiv$") the event $\{r, b\}$. When it is obvious that the draw is random, we may omit the word "random"; thus, here are two other ways to write the event:
        a drawn marble is red or blue;
        a red or blue marble is drawn.
Alternatively, we could write this same event this way:
        a red marble is drawn or a blue marble is drawn.

Note: In considering the occurrence of any specific event, the procedure is (imagined to be) run only once. In Procedure #6, only one marble is (imagined to be) drawn and its color is observed. The meaning of any event, including its occurrence, always goes back to the procedure for that sample space.

As another example, consider this event:
        a red marble is drawn and a blue marble is drawn.
This is the (an) impossible event, $\{\ \}$, for Procedure #6 since it is impossible to draw a marble that is both red and blue, or to draw a red marble and a blue marble on a single draw of one marble.

Whenever we write a statement defining an event, or an equivalent list, that statement may become true or false according to whether the event occurs or not, respectively. The methods of probability help us attach a probabilistic truth value to such events.

Simple Probability

In the study of logic, a statement is a sentence that can be judged true or false. In the study of probability, a statement can additionally be (or define) an event. Thus, an event may be expressed as a list of outcomes or as an equivalent statement, either of which has a probabilistic truth value. A probability value is a measure of relative knowledge, rather than absolute knowledge, about an event.

The *probability* that a specific event occurs (will occur) in a given circumstance (run) is the relative frequency with which that event would occur in a large number of repeated (real or imagined) identical circumstances (runs).

For the probability that an event A occurs ("occurs" is usually used instead of "will occur") in a single run of a particular procedure, we use the symbolism P(A), which is read "P of A".

$P(A)$ = the probability that event A occurs

(in a single run of the procedure).

This merely identifies the meaning of the symbolism in words; we have yet to clarify what numerical value is meant. From our definition of probability given above, we have this:

$P(A)$ = the relative frequency of the occurrence of A

(if we were to run, or imagine running, the procedure a large number of times in identical circumstances).

Thus, we have

$$P(A) = \frac{\text{the number of times A would occur}}{\text{the number of times S would occur}}$$

(in a large number of runs).

Remember that S occurs in each run, so the denominator is just the number of times the procedure is imagined to be run.

From this formula, it is easy to see that

$P(S) = 1$ and

$P(\{\ \}) = 0$

Other important implied features of any probability function on a sample space are given in the Glossary of Symbols.

Simple Probability Examples

As a simple example,
consider Procedure #6, for which $S \equiv \{r, g, b\}$.

One event is this: On a random draw the drawn marble is not red. (More simply we might say, "a drawn marble is not red"; we may assume that the action is to be accomplished in a random manner.) For ease of discussion, we may assign the symbol A to this event:
$A \equiv$ On a random draw, the drawn marble is not red.
Equivalently, $A \equiv \{b, g\}$ (the listing form of the event).

Let us do one run of Procedure #6. Suppose we draw a red marble, i.e., the outcome is red. Then event A did not occur. But in practical circumstances, we use probabilities when we have to make decisions about things before the actual outcome is known; before the run we need to find the value of P(A).

As suggested above, P(A) = the relative frequency with which A would occur in a large number of identical circumstances. Considering that there is 1 blue marble, 2 green marbles, and 5 red marbles in the urn, we can reason that the relative frequency with which we would not draw a red marble in a large number of identical circumstances is 3/8:

$$P(A) = P(\{b, g\}) = \frac{\text{the number of times A would occur}}{\text{the number of times S would occur}} = \frac{3}{8}$$

Similarly, suppose we let
$B \equiv$ a randomly drawn marble is red or blue $\equiv \{r, b\}$. Then

$$P(B) = P(\{r, b\}) = \frac{\text{the number of times B would occur}}{\text{the number of times S would occur}} = \frac{6}{8}$$

We might also write this:
a red marble is drawn or a blue marble is drawn.
This event is also equivalent to $\{r, b\}$:
$B \equiv$ a red marble is drawn or a blue marble is drawn $\equiv \{r, b\}$.

As another example, consider the event C, where we let
$C \equiv$ a red marble is drawn and a blue marble is drawn.
This is the (an) impossible event, { }: $C \equiv \{\ \}$
Remember that the meaning of any event always goes back to the procedure for that sample space. For Procedure #6, it is impossible to draw a marble that is both red and blue, or to draw a red marble and a blue marble on a single draw of one marble: $P(C) = P(\{\ \}) = 0$

Remember, whenever we write an event as a statement or an equivalent list, that statement may be true or false. The methods of probability help us attach a probabilistic truth value to such events.

Event Operations and Venn Diagrams

As soon as one or more ideas has been introduced in any subject, the human mind seeks to transform them. This may be as simple as negating a single idea or making operations between two ideas.

To further our understanding of probability and its power, we must define and understand some simple operations that can be performed with events on any given sample space S.
There are three such operations that are particularly useful: "complement", "or", and "and". These operations are defined and their symbols are given below.

A complement

The *complement of A*, also called *A complement*, is the event that contains all outcomes in S that are not in A.

$\overline{A}$ is the symbol we use for A complement.

A or B

The symbolic form $A \vee B$ can be read "A or B". $A \vee B$ is the event that contains all the outcomes that are in A or in B (or in both).
Notice that $A \vee B \equiv B \vee A$.

The use of the symbolic form is important since it helps us distinguish the use of "or" in ordinary circumstances from the particular meaning of the operation "or" between events.

A and B

The symbolic form $A \wedge B$ can be read "A and B". $A \wedge B$ is the event that contains all the outcomes that are both in A and in B.
Notice that $A \wedge B \equiv B \wedge A$.

The use of the symbolic form is important since it helps us distinguish the use of "and" in ordinary circumstances from the particular meaning of the operation "and" between events.

For each of these three operations, the resulting event is an event on the original sample space.

Event Operations and Venn Diagrams

S

S is the sample space.
A is an event on S.
The event A is shaded red.

S

not A is the *complement* of A, which contains all outcomes that are not in A.
The complement of A is shaded green.
$\overline{A}$ is the complement of A.
$\overline{A}$ is A complement.
$\overline{A}$ is an event on S.
The event $\overline{A}$ is shaded green.

S

$A \vee B$ is the event that contains all the outcomes that are in A or in B (or in both).
$A \vee B$ is an event on S.
The event $A \vee B$ is shaded blue.
$A \vee B \equiv B \vee A$.

S

$A \wedge B$ is the event that contains all the outcomes that are both in A and in B.
$A \wedge B$ is an event on S.
The event $A \wedge B$ is shaded.
$A \wedge B \equiv B \wedge A$

Complements Rule

Now that we see the Venn diagrams, and have a way of intuitively understanding the three simplest operations ("complement", "or", and "and") that we can carry out with events, we might ask how the probabilities of various events relate to these operations. The easiest operation is the complement operation, so we examine it first.

There are many probability problems in which it is easier to compute $P(\overline{A})$ than $P(A)$.

Fortunately, these two are simply related, which we can realize by looking at their Venn diagrams and noticing that $A \vee \overline{A} \equiv S$: since every outcome is either in A or in $\overline{A}$, every outcome is in $A \vee \overline{A}$; consequently, $S \equiv A \vee \overline{A}$.

It is important to note further that $A \wedge \overline{A} \equiv \{\}$, that is, $A \wedge \overline{A}$ has no outcomes in it (since no outcome can be both in A and in $\overline{A}$); this is because $\overline{A}$ has only outcomes that are not in A.

Intuitively, then it can be seen from the Venn diagrams of A and of $\overline{A}$, that since $P(S) = 1$,

$$1 = P(A) + P(\overline{A}) \qquad \text{(the } complements \text{ } rule\text{)}$$

The complements rule is often expressed in the equivalent forms:

$$P(A) = 1 - P(\overline{A}) \quad \text{or} \quad P(\overline{A}) = 1 - P(A)$$

As an example we return to Procedure #6, for which $S \equiv \{r, g, b\}$.
Again let $A \equiv \{b, g\}$ and $B \equiv \{r, b\}$.
Suppose we want to evaluate $P(A)$. (We still imagine only one marble is drawn.)
We previously computed $P(A)$, but if we had not already done so,
we could have written $\overline{A} \equiv \{r\}$, and easily computed $P(\overline{A}) = \dfrac{5}{8}$

Now we use the complements rule:
$$P(A) = 1 - P(\overline{A}) = 1 - \frac{5}{8} = \frac{8}{8} - \frac{5}{8} = \frac{3}{8}, \text{ which we got previously.}$$

In this example, there is no apparent advantage to computing $P(A)$ directly, as we did previously, or to using the complements rule. On the other hand, there are many problems where it is extremely advantageous to use the complements rule. (See Batch Testing in the Practical Applications.)

Addition Rule

Again turning to the Venn diagrams, we can intuitively understand how the probabilities of A, B, $A \vee B$, and $A \wedge B$ are related to one another:

$$P(A) + P(B) = P(A \vee B) + P(A \wedge B), \quad \text{which is equivalent to}$$

$$P(A \vee B) = P(A) + P(B) - P(A \wedge B) \qquad \text{(the } \textit{addition rule}\text{)}$$

When we compute $P(A \vee B)$ the reason we must subtract $P(A \wedge B)$ is that it is included in each of $P(A)$ and $P(B)$, and we only want to count it once.

As an example we return to procedure #6, for which $S \equiv \{r, g, b\}$.
Again let $A \equiv \{b, g\}$ and $B \equiv \{r, b\}$.
Suppose we want to evaluate $P(A \vee B)$.
(We still imagine only one marble is drawn.)
We use the addition rule. We previously found $P(A) = \frac{3}{8}$ and $P(B) = \frac{6}{8}$.

Since $A \wedge B \equiv \{b\}$, we also see that $P(A \wedge B) = \frac{1}{8}$.

Thus, $P(A \vee B) = P(A) + P(B) - P(A \wedge B) = \frac{3}{8} + \frac{6}{8} - \frac{1}{8} = \frac{8}{8} = 1$

The result $P(A \vee B) = 1$ is no surprise if we noticed that $A \vee B \equiv \{r, g, b\} \equiv S$, since we know that $P(S) = 1$.
Thus, we have some evidence, in this simple example, that the addition rule is valid.

For a more informative example, we return to Procedure #1, where $S \equiv \{1, 2, 3, 4\}$.
Let $A \equiv \{2, 3\}$ and $B \equiv \{3, 4\}$.
As in the previous example, it is easiest to evaluate $P(A \vee B)$ directly, that is, without the addition rule:
$P(A \vee B) = P(\{2, 3, 4\}) = \frac{3}{4}$ <u>assuming</u> each side of the die is equally likely.
However, for the purpose of illustrating the addition rule, let us use it here to evaluate $P(A \vee B)$ again.
First, we need $P(A)$, $P(B)$, and $P(A \wedge B)$:
$P(A) = P(\{2, 3\}) = \frac{2}{4}$, $P(B) = P(\{3, 4\}) = \frac{2}{4}$, and $P(A \wedge B) = P(\{3\}) = \frac{1}{4}$

Then we have $P(A \vee B) = P(A) + P(B) - P(A \wedge B) = \frac{2}{4} + \frac{2}{4} - \frac{1}{4} = \frac{3}{4}$,
which is the result we obtained above.

This result is further evidence that the addition rule works.

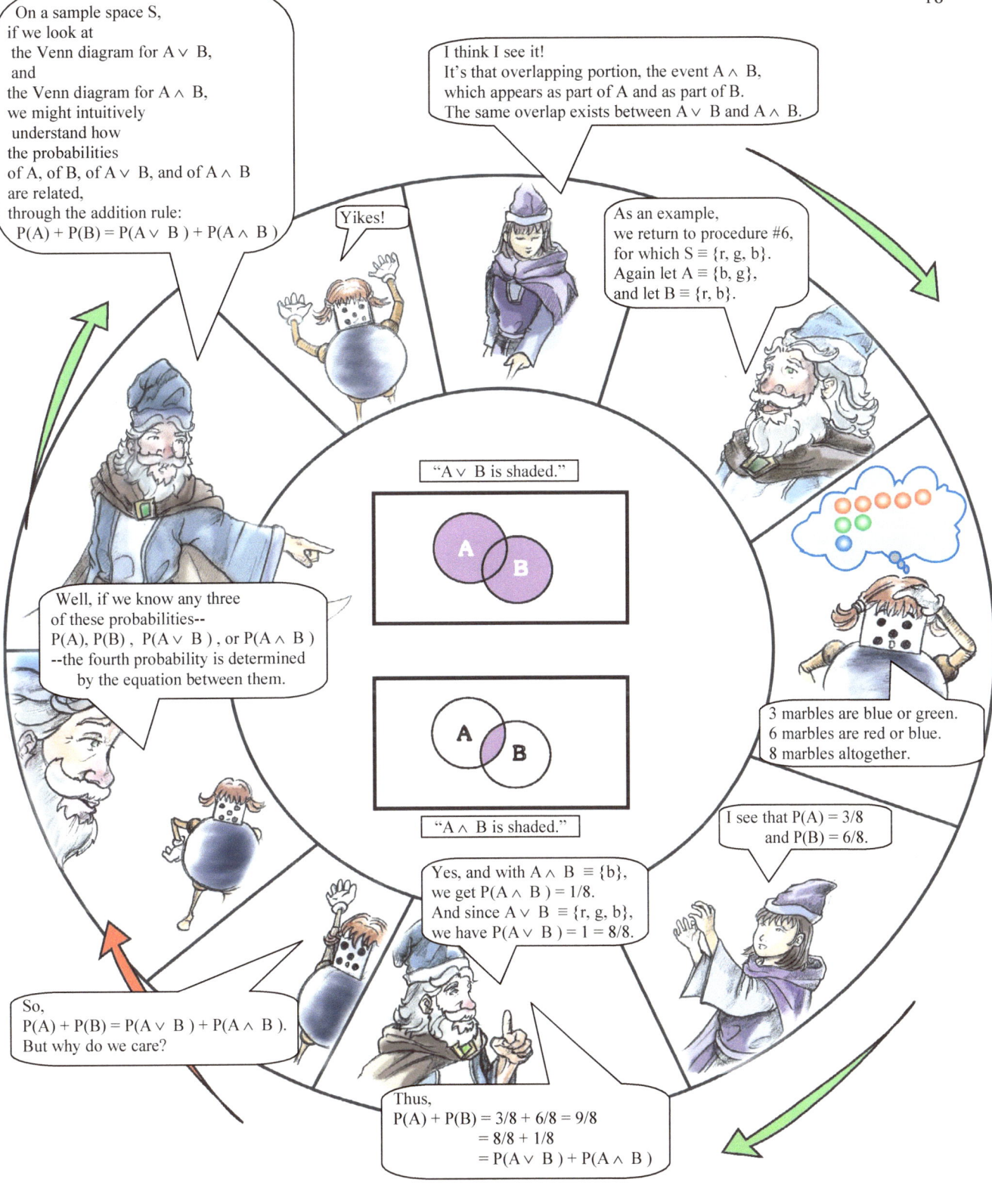

Conditional Probability

A potent concept in probability is *conditional probability*.

It is easiest to discuss conditional probability in symbolic form:
    P(A|B) is read "P of A given B",
        or more fully, "the probability that event A occurs given that event B occurs".

At its simplest level, this symbolism expresses the probability of event A occurring given that event B is assumed to occur at the same time. In other applications, B may be assumed to occur at a different time than that for which we are considering, or speculating, as to whether A will occur; the occurrence of B might be before or after the occurrence of A. In all cases, the occurrence of A is in the future, and that of B is at least hypothetical (assumed) and is sometimes actual.

For example, we might ask what is the probability that a person will develop lung cancer within ten years if it is assumed that the person maintains a pack-per-day smoking habit. Both events are hypothetical. What is the probability that Endeavor will beat Starling in today's race given that Starling is lame? If Starling is already universally known to be lame, there may be no benefit in writing a conditional probability statement; we may simply say P(Endeavor will win) = 1. On the other hand, if we wish to make explicit the effect of Starling's known lameness, or if Starling's lameness is hypothetical, we have a conditional probability. For another example, we may wish to evaluate P({b, g}|{r, b}) for events on S for Procedure #6.

Generally, how can we evaluate P(A|B)? If A and B are events on the same sample space, we can use relative frequency, and hence use the ratio of numbers of times events can occur in many runs of the procedure.

$$P(A|B) = \frac{\text{the number of times } A \wedge B \text{ would occur}}{\text{the number of times } B \text{ would occur}}$$

We have the option of using an equivalent alternative formula:

$$P(A|B) = \frac{P(A \wedge B)}{P(B)}$$

As an example we return to Procedure #6, for which S ≡ {r, g, b}. Again let A ≡ {b, g} and B ≡ {r, b}. Suppose we want to evaluate P(A|B). (We still imagine only one marble is drawn.) We use $P(A|B) = \frac{P(A \wedge B)}{P(B)}$.

Since A ∧ B ≡ {b}, we get $P(A \wedge B) = \frac{1}{8}$. We previously found $P(B) = \frac{6}{8}$.

Thus, $P(A|B) = \frac{1/8}{6/8} = \frac{1}{6}$.

The Product Rule

The conditional probability formula $P(A|B) = \dfrac{P(A \wedge B)}{P(B)}$ implies the *product rule*:

$$P(A \wedge B) = P(A|B)P(B) = P(B|A)P(A) \qquad \text{(Product rule)}$$

We continue to imagine the future in our concept of something that will happen, or an event that will occur, regardless of the usage of "is" instead of "will be", or "event A occurs" instead of "event A will occur". It is with conditional probability and the product rule that we realize there are many situations in which we can gain additional power if we distinguish more carefully between past, present, and future.

We developed the conditional probability formula for the case that A and B are events on the same sample space; however, in many interesting situations they are not. Nonetheless, in evaluating $P(A \wedge B)$ and $P(B)$, both events A and B are speculative, to occur or not occur in the future. Furthermore, as mentioned previously, B may occur before or after A occurs. In these cases, $A \wedge B$ may be interpretted as the event "A after B", "B before A", etc.; likewise, we may have "B after A", etc.

Regardless of how A and B are temporally related,
$P(A \wedge B)$ is still the probability that A occurs and B occurs;
$P(A|B)$ is the probability that A occurs given that B is assumed to occur.

The product rule is especially useful in determining *independence* of events.
Events A and B are *independent* if and only if $P(A \wedge B) = P(A)P(B)$,
which is true if and only if $P(A|B) = P(A)$,
which is true if and only if $P(B|A) = P(B)$.

For our example of $A \equiv \{b, g\}$ and $B \equiv \{r, b\}$ from Procedure #6,
we previously found $P(A) = \dfrac{3}{8}$ and $P(B) = \dfrac{6}{8}$.

So $P(A)P(B) = \dfrac{3}{8} \bullet \dfrac{6}{8} = \dfrac{3 \bullet 6}{8 \bullet 8} = \dfrac{18}{64}$.

We also previously found $P(A \wedge B) = \dfrac{1}{8}$.

So we see that $P(A \wedge B) \ne P(A)P(B)$.

Thus, A and B are *dependent*, not independent.
Alternatively, we already knew that $P(A|B) = \dfrac{1}{6}$, which is not equal to $P(A) = \dfrac{3}{8}$; again we conclude that A and B are not independent events; they are dependent events.

Decision Trees

A *decision tree* is often useful when the procedure is complex. The decision tree can help us construct the sample space and its associated probabilities.

As an example let us consider Procedure #8: An urn contains 3 red marbles, 1 green marble, and 2 blue marbles, all of which differ from one another only in color; from the urn we successively draw a first marble and then a second marble, without having replaced the first marble, and observe the ordered-pair of colors of the first and second marbles drawn.

The decision tree for this procedure looks like this:

```
                                          ordered-pair            probability
    1st draw           2nd draw             outcome

                              r              rr            (3/6)(2/5) = 6/30
              r               g              rg            (3/6)(1/5) = 3/30
                              b              rb            (3/6)(2/5) = 6/30

                              r              gr            (1/6)(3/5) = 3/30
              g
                              b              gb            (1/6)(2/5) = 2/30

                              r              br            (2/6)(3/5) = 6/30
              b               g              bg            (2/6)(1/5) = 2/30
                              b              bb            (2/6)(1/5) = 2/30
```

The probabilities for the second draw are affected by the first draw.
The product rule gives the probabilities of each of the ordered-pair outcomes.
Notice that the sum of these final probabilities is 30/30 = 1.

The sample space is this: S ≡ {rr, rg, rb, gr, gb, br, bg, bb}.

Let A ≡ a randomly chosen ordered-pair has a green marble.
Then A ≡ {rg, gr, gb, bg}.

P(A) = P({rg}) + P({gr}) + P({gb}) + P({bg})
     = 3/30 + 3/30 + 2/30 + 2/30   = 10/30

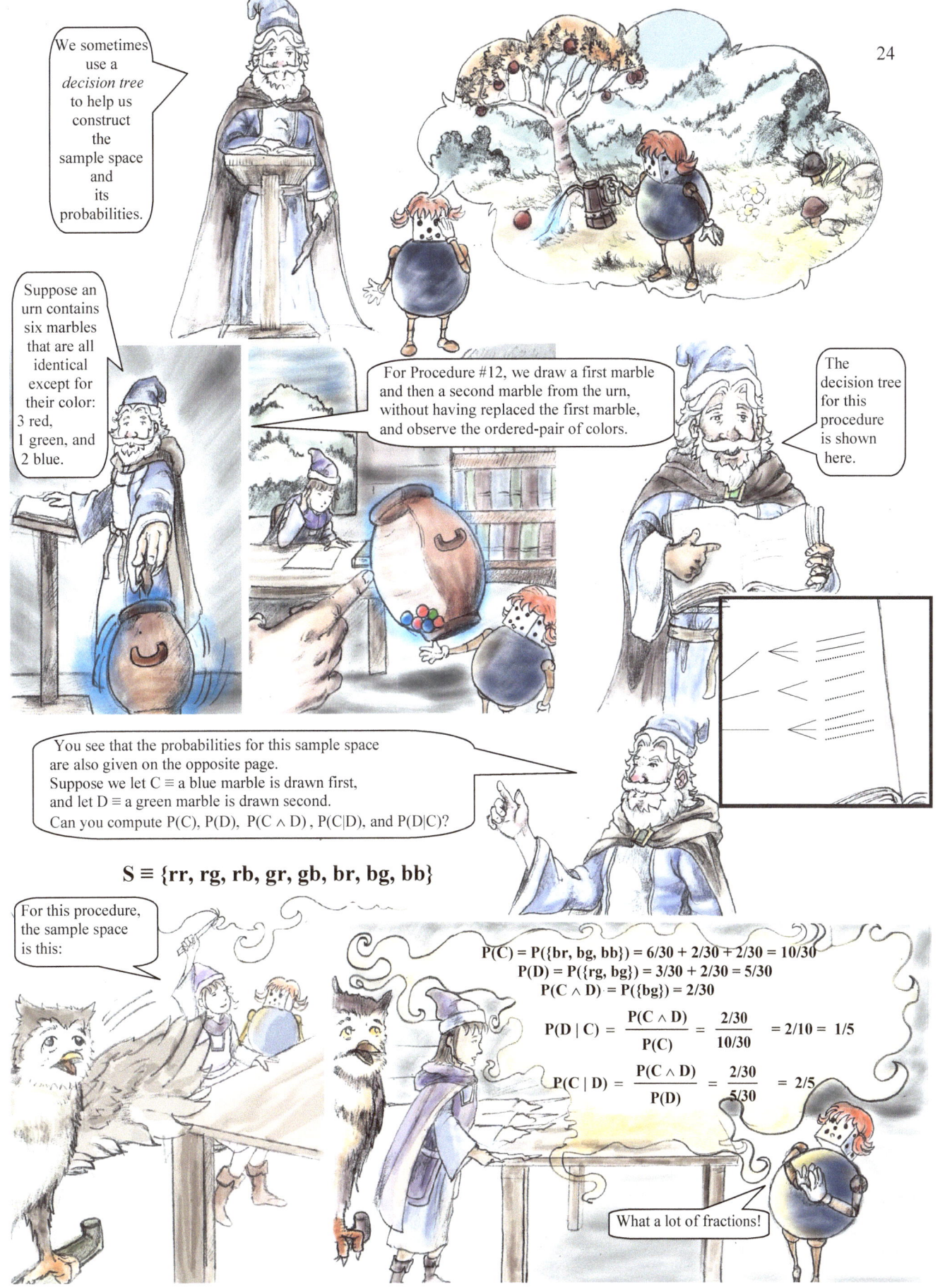

Practical Applications

1. Tossing a Coin; Playing a Slot Machine
(an example of simple probability)

To improve our understanding of the meaning of probability, we may examine the toss of a coin in detail.

Let us define two events to examine.
$A \equiv$ a head comes up on the second toss of the coin
$B \equiv$ a tail comes on the first toss of the coin

If we have examined a coin and believe it to be a fair coin, i.e., one for which heads and tails are equally likely, then we may write P(we get head on a single toss of the coin) = 1/2. Particularly, $P(A) = 1/2$

Some people find it difficult to accept that

$P(A|B) = P(A) = 1/2$

Among some of us there seems to be an unspoken sense that some probability god keeps track of how many heads have been tossed and that he/she will even everything out. Rather, it is large numbers of tosses, and nothing else, that will even everything out. If the coin is fair, then from our definition of probability in terms of relative frequency,

$$P(A) = \frac{\text{the number of times A would occur}}{\text{the number of times S would occur}} = 1/2$$

(for a large number of runs)

On the other hand, if we never examined the coin before we began tossing it, and then we get many consecutive heads, we may rightly begin to suspect that the coin has only an obverse (heads only).

If a coin is not fair because, for example, it has no tail side, eventually we would become convinced that $P(A) = 1$

This same logic must be applied to playing slot machines. If a slot machine has been programmed to give a 97% return rate, that is what dictates the next play regardless of the number of times the machine has not paid. Any other conclusion must be equated to a challenge of whether the posted return rate is correct.

Whether tossing a coin or playing a slot machine, the next play is independent of all past outcomes.

2. Cancer in Cell Phone Users
(an example of conditional probability)

Do cell phone users run a higher risk than normal of developing cancer of the brain or nervous system?

In a recent study in Denmark, it was found that in a group of 420,000 cell phone users 135 people developed cancer of the brain or nervous system. Based on this data, what is the probability that you will develop either of these cancers if you are a cell phone user?

We may define the procedure this way: Randomly choose a person from the population of Denmark, and check as to whether or not the person is and continues to be a cell phone user, and, in the future, for all those who continue their use of cell phones we will check whether or not the person is diagnosed with cancer of the brain or nervous system. (In this specification of the procedure, we assume that every cell phone user will be tested for the cancer.)

Let $A \equiv$ a (randomly chosen) person will develop cancer of the brain or nervous system.

Let $B \equiv$ a (randomly chosen) person is and continues to be a cell phone user.

P(a person will develop cancer of the brain or nervous system if that person is a cell phone user)

$$= P(A|B)$$

Using the formula for conditional probability,

$$P(A|B) = \frac{\text{the number of times } A \wedge B \text{ would occur}}{\text{the number of times } B \text{ would occur}}$$

Thus, the data from the Denmark study imply that

$$P(A|B) = \frac{135}{420,000} = 0.000321 \quad \text{(approximately)}$$

It also known from other studies that

$$P(A) = 0.000340 \quad \text{(without regard for cell phone use)}$$

At first glance one might wonder if cell phone use inhibits the cancer; however, these two probabilities are close enough that nothing can be concluded about the benefit or detriment of cell phone use to your health. (The methods of inferential statistics resolve this issue.)

3. Lotteries and Sweepstakes
(an example of expected value)

How much is a lottery ticket worth?

Most of us who play games of chance (gambling) have an intellectual awareness that the game is a losing game. It is the excitement of the uncertainty of the game, the possibility of winning or losing, rather than the fairness of the game, that entices us. For that pleasure, we are willing to pay money; it is simply a kind of entertainment. In Las Vegas the activities and results might be more immediate, but theoretically they are no different than lotteries or sweepstakes. Also, something that is often overlooked by those who are eager to play such games (or to justify their involvement) is that any proposed winnings of a game or gaming strategy must take into account the relevant taxation of winnings.

One of the additional issues with lotteries and sweepstakes is that they often quote the prize as a large amount payable over a twenty-year period. That the bankable value of such prizes is much lower is clarified by the (optional) immediate cash payout that is sometimes offered, which is approximately half as much as the total twenty-year amount to be paid.

Expected value is the name given to the value of such games of chance. We use the symbol E for expected value, and we define it this way: $E = \sum_i [x_i P(\{x_i\})]$

A typical probability for winning a lottery is 1/(100 million). Imagine we pay $1 to play to win a $5 million immediate cash payout. What is the bankable value of participating?

Let $A \equiv$ we win. Then, $\overline{A} \equiv$ we lose. The $1 is paid whether we win or lose; so, if we win, our net winning amount is $5 million - $1 = $4,999,999.
The sample space is $S \equiv \{win, lose\} \equiv \{\$4,999,999, -\$1\}$

$$E = \sum_i [x_i P(\{x_i\})]$$

$= \$4,999,999[1/(100 \text{ million})] + (-\$1)[99,999,999/(100 \text{ million})]$

$\approx \$00.05 - \$1.00 = -\$00.95 = -95$ cents.

So (ignoring the taxation of winnings) the value of the ticket before the drawing is 5 cents; you pay $1 for it. How much such excitement can you afford? Go to it! Good luck!

(If the immediate payout to a single winning ticket is $60 million, then the computation yields this: $E \approx -40$ cents, assuming the probabilities remain the same. So, before the drawing, the $1 ticket is worth 60 cents.)

4. Batch Testing
(an example of the complements rule)

A common method of testing, to either accept or reject a production batch, is to check a relatively small sample, and base the acceptance of the whole batch on the sample.

Suppose 5,000 light bulbs have been produced. Before the expense of packaging we randomly choose a sample of ten light bulbs and check them for faults. If one or more of the 10 bulbs is defective, we will reject the whole batch of 5,000. If the production defect rate has usually been running at 2%, what is the probability that we will reject the whole batch?

Because the batch is so large and the sample so relatively small, we have approximate independence of sampling. In other words, though we take the sample without replacement of any bulbs, we can treat the sample as though the bulbs were chosen independently. What this means is that the likelihood of any given sample bulb being defective is approximately unrelated to the quality of any other sample bulb.

Let $A \equiv$ at least one sample bulb is defective

Then $\overline{A} \equiv$ all sample bulbs are good

P(we reject the whole batch) = P(A)

P(A) is laborious to compute directly, but it is trivial to compute $P(\overline{A})$.

$P(\overline{A})$ = P(all sample bulbs are good) = $(98\%)^{10} = (0.98)^{10} = 0.817$

P(we reject the whole batch) = P(A) = $1 - P(\overline{A}) = 1 - 0.817 = 0.183$

P(we reject the whole batch) = 0.183

Notice that for this same defect rate, 2%, the probability that the whole batch will be accepted is this:

P(we accept the whole batch) = 0.817

If the defect rate for this batch is actually 5%, what is the probability that the whole batch will be rejected?

This same method can be used to reduce the cost of expensive medical tests.

5. Interrelated Manufacturing Defects
(an example of independence/dependence)

Imagine that Fast Track Computers (FTC) has purchased a batch of twenty rejected computers from a major manufacturer in hopes of easily correcting the problems and selling at a profit.

These twenty computers have been numbered (1-20), tested, and labelled for the type of fault they contain:
4 of the computers have both a defective motherboard and a defective video card (dmdv);
for 5 of the computers, only the motherboard is defective (dmo);
for 2 of the computers, only the video card is defective (dvo).

One of the FTC engineers has suggested that there is a relationship between a computer having a defective motherboard and a computer having a defective video card. Is he right, or are these issues independent of one another?

(1) First analysis

To do a probabilistic analysis we begin by defining the relevant events:
let $A \equiv$ a (randomly chosen) computer has a defective motherboard;
let $B \equiv$ a (randomly chosen) computer has a defective video card.

Next we define an imagined procedure and sample space. We have more than one choice for the procedure, but one way is this: We randomly choose a computer, and observe its number and whether it is ok, dmo, dvo, or dmdv.

For this procedure we have the following sample space:
$S \equiv \{$(1,dmo), (2,dmo), (3,dmo), (4,dmo), (5,dmo), (6,dvo), (7,dvo), (8,dmdv), (9,dmdv), (10,dmdv), (11,dmdv), (12,ok), (13,ok), (14,ok), (15,ok), (16,ok), (17,ok), (18,ok), (19,ok), (20,ok)$\}$

$A \equiv \{$(1,dmo), (2,dmo), (3,dmo), (4,dmo), (5,dmo), (8,dmdv), (9,dmdv), (10,dmdv), (11,dmdv)$\}$
$B \equiv \{$(6,dvo), (7,dvo), (8,dmdv), (9,dmdv), (10,dmdv), (11,dmdv)$\}$
Then, $A \wedge B \equiv \{$(8,dmdv), (9,dmdv), (10,dmdv), (11,dmdv)$\}$

A and B are independent iff $P(A \wedge B) = P(A)P(B)$.
Let us check: $P(A) = 9/20$, and $P(B) = 6/20$,
so $P(A)P(B) = (9/20)(6/20) = 54/400 = 27/200$;
however, $P(A \wedge B) = 4/20 = 1/5$. Thus, $P(A \wedge B) \neq P(A)P(B)$
So, the events A and B are dependent.

(2) <u>Second analysis</u>

Notice that we could have used an alternative procedure. Suppose the procedure is this: we randomly choose a computer, and observe whether it is ok, dmo, dvo, or dmdv. (We ignore its number.)

For this procedure we have the following sample space:
S ≡ {dmo, dvo, dmdv, ok}

The advantage of using this sample space is that there are fewer outcomes. The disadvantage is that the outcomes are not equiprobable.

The statements defining each of A and B are the same as before, but on this sample space the listed forms of A and B are different than before:
A ≡ {dmo, dmdv}
B ≡ {dvo, dmdv}
A ∧ B ≡ {dmdv}

The rest of the analysis remains the same, with P(A), P(B), and P(A ∧ B) being computed exactly as above based on the computer data. Thus, we still conclude that
P(A ∧ B) ≠ P(A)P(B), that is, the events A and B are dependent.

Glossary of Terms

Addition Rule: $P(A) + P(B) = P(A \vee B) + P(A \wedge B)$

Certain event: An event (on S) that occurs whenever the procedure is run is a certain event.

Complement: The complement of event A on S is the event containing all the outcomes in S that are not in event A.

Complements Rule: $P(A) + P(\overline{A}) = 1$

Conditional probability is the probability that a named event occurs given that some other named event occurs.

Decision tree: A decision tree provides a branching structure that aids in the organization and tracking of potentially non trivial outcomes of a complex procedure.

Dependent events: Two events are dependent if and only if they are not independent.

Disjoint events: Two events are disjoint if and only if they have no outcomes in common.

Equiprobable sample space: An equiprobable sample space is one in which each outcome is equally likely.

Equivalent events: Two events are equivalent on some sample space if and only if they always occur or do not occur simultaneously in a run of the procedure.

Expected value is the average of the numerical outcomes of a procedure run a large number of times.

Event: An event on some sample space is a complete or partial list (set) of possible outcomes belonging to that sample space.

Event operation: An event operation is a logical operation on an event or between events: "complement," "or," and "and" are the operations introduced in this book.

Experiment: A probability experiment is the run or running of a procedure one or more times.

Iff: "iff" means "if and only if"

Impossible event: An event (on S) that cannot occur when the procedure is run.

Independent events: Two events are independent iff the occurrence of one event has no impact on the probability of the occurrence of the other event.

Mutually exclusive: Mutually exclusive means the same thing as disjoint.

Occur: An event is said to occur if on a run of the procedure the outcome of that run is contained in the event.

Outcome: An outcome is the result that is observed, or could be observed, according to the procedure.

Probability: Probability is the relative frequency with which something would happen in a large number of identical circumstances. It is a measure of relative knowledge.

Procedure: A probability procedure specifies an action and an observation.

Product Rule: $P(A \wedge B) = P(A|B)P(B)$

Relative frequency: In a repetitive activity, such as running a probability procedure, the relative frequency of a given event is the quotient of the number of times the given event occurs and the number of times the procedure is run:
$$\text{Relative frequency of a given event} = \frac{\text{number of times the given event occurs}}{\text{number of times the procedure is run}}.$$
Notice that the relative frequency is meaningful whether or not we run the procedure many times.

Run: A procedure is run when the specified action and observation are carried out.

Sample space: The sample space associated with a procedure is the list (set) of all outcomes for that procedure.

Statement: A statement is a sentence that can be judged true or false. In probability, a statement can be an event, and has a probabilistic truth value.

Venn diagram: A Venn diagram is a picture that represents a sample space and an event or an operation between events on the sample space.

Glossary of Symbols

A, B, C, etc.   Roman capital letters are used to represent events on a sample space.

$A \wedge B$ This is the event that contains every outcome that is both in A and in B. It is read "A and B".

$A \vee B$ This is the event that contains every outcome that is in A or in B, or in both. It is read "A or B".

$\overline{A}$ $\overline{A}$ is the complement of A. It contains all outcomes that are not in A. It is an event that occurs when the procedure is run and A does not occur.

$\{\ \}$ $\{\ \} \equiv$ the impossible event

$=$ This is the symbol for equality between two numbers.

$\equiv$ This is the symbol for equivalence between two events.

E   Expected value $= E = \sum_{i}[x_i P(\{x_i\})]$

P(A)   P(A) is the probability that A occurs on a single run of the procedure.

$$P(A) = \frac{\text{the number of times A would occur}}{\text{the number of times S would occur}}$$
(in a large number of runs)

Note that we always have $0 \leq P(A) \leq 1$; note also that $P(S) = 1$ and $P(\{\ \}) = 0$.

P(A|B)   This is the symbol for the conditional probability that A occurs given that B occurs.

S $S \equiv$ sample space. S contains all possible outcomes of a given procedure.

$\sum_{x_i \text{ in } A} P(\{x_i\})$ This symbol is the sum of all values $P(\{x_i\})$ where the x_i are outcomes contained in event A on the sample space. Note that we always have

$$P(A) = \sum_{x_i \text{ in } A} P(\{x_i\}) .$$

x_i x_i is a possible outcome of a procedure. It might be a number, the side of a coin, the color of a marble, an ordered pair, etc. The subscript, i, is a way of naming the various possible outcomes in a list.

Appendix

Fractions are fundamental to probability.

Some people find fractions difficult to master, but when we are using small denominators like 4, 6, 8, etc., the fractions are all very simple. Also, operations between fractions remain simple when the fractions are simple.

First, remember that a fraction is a number $\frac{a}{b}$ that has a numerator (above the fraction bar) and denominator (below the fraction bar). In probability the numerator must be a whole number: 0, 1, 2, 3, ... The denominator must be a natural number: 1, 2, 3, ...

Examples: $\frac{1}{4}$, which is read "one fourth"; $\frac{3}{4}$, which is read "three fourths"; $\frac{5}{8}$, which is read "five eighths"; etc.

Remember too that $1 = \frac{2}{2} = \frac{4}{4} = \frac{8}{8} = \frac{6}{6} = \frac{5}{5}$ and so forth.

In probability there are four operations between fractions: addition (+), subtraction (-), multiplication (×), and division (÷). The following shows the very simple ideas that are needed and used in the text to allow an initial appreciation and understanding of probability.

For addition of fractions we have $\frac{a}{c} + \frac{b}{c} = \frac{a+b}{c}$.

Examples: $\frac{1}{4} + \frac{2}{4} = \frac{3}{4}$, $\frac{2}{8} + \frac{5}{8} = \frac{7}{8}$, $\frac{2}{8} + \frac{4}{8} = \frac{6}{8}$

A similar rule, and just as simple, is the rule for subtraction of fractions:

$$\frac{a}{c} - \frac{b}{c} = \frac{a-b}{c}$$

Examples: $\frac{3}{4} - \frac{2}{4} = \frac{1}{4}$, $\frac{7}{8} - \frac{2}{8} = \frac{5}{8}$, $\frac{7}{8} - \frac{3}{8} = \frac{4}{8}$

Our examples are simple because quite often in probability the denominators in a given problem are identical; furthermore, we usually do not need to simplify a fraction (reduce it) when we finish adding up the fractions. On the other hand, if the need to build or reduce a fraction, that is, to change the denominator, should arise, the following is the rule:

$$\frac{a}{b} = \frac{a \times c}{b \times c}$$

Examples: $\dfrac{3}{4} = \dfrac{3 \times 2}{4 \times 2} = \dfrac{6}{8}$, $\dfrac{2}{3} = \dfrac{2 \times 2}{3 \times 2} = \dfrac{4}{6}$, $\dfrac{4}{8} = \dfrac{1 \times 4}{2 \times 4} = \dfrac{1}{2}$

Note that in the first two examples we built up the fraction; that is, we made the denominator larger. In the third example we reduced the fraction; that is, we made the denominator smaller.

We should also remember the rule for multiplication of fractions:

$$\frac{a}{b} \times \frac{c}{d} = \frac{a \times c}{b \times d}$$

Examples: $\dfrac{2}{3} \times \dfrac{4}{5} = \dfrac{2 \times 4}{3 \times 5} = \dfrac{8}{15}$, $\dfrac{3}{4} \times \dfrac{1}{6} = \dfrac{3 \times 1}{4 \times 6} = \dfrac{3}{24} = \dfrac{1}{8}$

Finally, we should remember the rule for division of fractions:

$$\frac{a}{b} \div \frac{c}{d} = \frac{a}{b} \times \frac{d}{c} = \frac{a \times d}{b \times c}$$

Examples: $\dfrac{2}{3} \div \dfrac{4}{5} = \dfrac{2 \times 5}{3 \times 4} = \dfrac{10}{12} = \dfrac{5}{6}$, $\dfrac{1}{4} \div \dfrac{5}{6} = \dfrac{1 \times 6}{4 \times 5} = \dfrac{6}{20} = \dfrac{3}{10}$

Operations in Set Theory and Logic

We beg the indulgence of the more advanced student for the choice of names of operations and their symbols in this text. In logic, to operate on or between statements, we have the operations of negation, disjunction (or), and conjunction (and), for which it is common to use the symbols $\sim$, $\vee$, and $\wedge$, respectively. In set theory, the analogous operations are complement, union, and intersection, for which the respective symbols are $\overline{}$, $\cup$, and $\cap$. For example, "A union B" can be written $A \cup B$.

In this text we have adopted the hybrid usage of $\overline{}$, $\vee$, and $\wedge$ for the operations between events, for which we use the words "complement," "or," and "and," respectively. We see that this is reasonable by referring to "Lists, Statements, and the Meaning of Events." There we see that an event can always be viewed as a list of outcomes, i.e., as a set, a capital letter (A, B, or C), or a statement. Either of the former two can be viewed as a symbol for the latter. For the example given in "Lists, Statements, and the Meaning of Events," we had {r, b} is equivalent to ($\equiv$) a randomly drawn marble is red or blue. Thus, given the possible justifiable choices to be made, the names and symbols for the operations were chosen to be the least confusing for someone who might never have seen them before in a mathematical context.

Index

Addition Rule: 17, 18, 32

Batch testing: ix, 29

Certain event: 5, 6, 32

Complement: 13, 14, 15, 16, 32

Complements Rule: 15, 16, 32

Compound event: 5

Conditional probability: 19, 20, 21, 22, 32

Decision tree: 23, 24, 32

Dependent events: 21, 22, 30, 31, 32

Disjoint events: 32

Equiprobable sample space: 31, 32

Equivalence: 3, 4, 20

Equivalent events: 3, 32

Expected value: 28, 32

Event: 5, 6, 7, 8, most pages

Event operations: 13-18, 32

Experiment: 32

Fractions: xiii, xiv, 24, 35, 36

Iff: 22, 32

Impossible event: 5, 6, 32

Independent events: 21, 22, 30, 32

Magic: xi

Modeling: x

Mutually exclusive: 33

Occur, occurrence: 7, 8, 19, 20, 33

Operations: 13-18

Outcome: 1, 2, 3, 4, 33

Probability: 9, 10, and most pages

Procedure: 1, 2, and most pages

Product Rule: 21, 22, 33

Relative frequency: 9, 10, 19, 20, 33

Run: 1, 2, 3, 4, 5, 7, 8, 19, 20, 33

Sample space: 3, 4, 5, 6, and most pages

Simple event: 5

Symbol, symbolism: xiii, 5, 7, 20

Statement: 7, 8, 33

Venn diagram: 13, 14, 18, 33

Puzzles for the Advanced Student

An advanced student may wish to test their comprehensive knowledge with following problems.

1. Define a procedure, sample space, probability function on S, and two events, A and B, such that A and B are not disjoint (i.e., $A \wedge B$ is not equivalent to { }) and A and B are independent.

There are many solutions to this problem.

Here is one solution. Let the procedure be that we roll a fair 6-sided die and observe the number that comes up. In this case we have the sample space, $S \equiv \{1,2,3,4,5,6\}$.
The probability function is defined by $P(S) = 1$, $P(\{ \}) = 0$,
$P(\{1\}) = 1/6$, $P(\{2\}) = 1/6$, $P(\{3\}) = 1/6$, $P(\{4\}) = 1/6$, $P(\{5\}) = 1/6$, $P(\{6\}) = 1/6$.
We now define $A \equiv \{1,2\}$ and define $B \equiv \{1,3,5\}$
Thus, we have A and B are not disjoint, i.e., we have $A \wedge B \equiv \{1\}$,
and we see that $P(A|B) = P(A)$ since we have $P(A) = \dfrac{1}{3}$ and $P(A|B) = \dfrac{P(A \wedge B)}{P(B)} = \dfrac{1/6}{3/6} = \dfrac{1}{3}$

2. If you were not able to solve Problem 1 before you read the solution, try to find another solution using the same procedure as given in the above solution, but defining a different A and a different B.

3. Once you can do Problems 2, try to find a solution to Problem 1 but for which you define a different procedure and the sample space is again equiprobable.

4. Once you can do Problem 3, try to find a solution to Problem 1, but for which the sample space is not equiprobable. (Try something simple to start. For example, look for a procedure that gives a sample space with only three outcomes. While noting that you could choose a trivial solution in which B is set equivalent to S, make your goal more creative than that.)

One solution to Problem 4 is this. Let the procedure to be that we flick a spinner in which there are four colored regions, red, blue, green, and yellow, of four different sizes, and we observe the color to which the spinner points after it comes to rest.
In this case we have the sample space, $S \equiv \{r,b,g,y\}$.
The probability function could be defined by $P(S) = 1$, $P(\{ \}) = 0$, for which there is no freedom of choice, and by the remaining four equalities:
$P(\{r\}) = 1/4$, $P(\{b\}) = 1/12$, $P(\{g\}) = 1/6$, and $P(\{y\}) = 1/2$, which add up to 1.
Now we define $A \equiv \{r,b\}$ and define $B \equiv \{b,g\}$, which are not disjoint, since $A \wedge B \equiv \{b\}$.

We see that $P(A) = 4/12$ and $P(B) = 3/12$, and so we have $P(A)P(B) = \dfrac{4}{12} \bullet \dfrac{3}{12} = \dfrac{1}{12}$

Also, we see that $P(A \wedge B) = P(\{b\}) = 1/12$.
Thus, we have that $P(A \wedge B) = P(A)P(B)$, so that A and B are independent events.

About the author:

Lin Sten has written fiction and nonfiction for many years. Paragon House published his nonfiction *Souls, Slavery, and Survival in the Molenotech Age* in 1999. He is currently working on several fiction and nonfiction projects while teaching mathematics and physics part-time.

He has taught probability in courses ranging over arithmetic, mathematical literacy, and statistics. He has often used probability as a systems analyst in the aerospace industry and to solve problems in physics.

He is intrigued by the mystery of nature and awed by the magical power of symbolism. The ways in which people overcome their phobias in mathematics and physics continually fascinate him, and everyday in the classroom he learns something more about how to help them. This book arose from materials especially prepared by him for students who had read but not sufficiently understood their textbooks.

He lives in California on the coast of the Pacific Ocean with his mate, Liliana, without whom this book would not have been written.

www.ingramcontent.com/pod-product-compliance
Lightning Source LLC
Chambersburg PA
CBHW051052180526
45172CB00002B/605